Earth
Our Big Home!

Om
KiDZ
An imprint of Om Books International

Bob was in his classroom, sitting next to his friend Nina. He was looking forward to his science class. Bob loved studying science. "Good Morning!" Miss Kelly greeted the children with a smile. "Today, we are going to learn about the Earth."

"Yes!" exclaimed Bob, happily.

"What is so interesting about the Earth? Why are you so excited?" Nina asked him.

Miss Kelly who had overheard their conversation smiled and turned to the class, "Can anyone tell me what is so special about our planet Earth?" she asked them.

Bob's hand shot up in the air. He seemed to have an answer to everything.

"Earth is a planet. It is the only planet where humans live!"

"What is a planet, Miss?" Nina asked.

Miss Kelly showed them a colourful model of the solar system. She asked all the children to gather around her table.

"Planets are large, natural objects that travel around stars," explained Miss Kelly, pointing towards the model. "The Sun is a star and eight planets move around it."

"Miss, why is Earth the only planet in the solar system that supports life?" asked Mary.

"That's because the other planets are either too hot or too cold. Earth, on the other hand, has an atmosphere that supports the right temperature or else we would all have been burnt or frozen," Bob said.

"That's right, Bob! This blue ball here is Earth. Just like the ball, Earth is round. As 71% of Earth is covered with water, it looks blue from space. Earth is the only planet that has oxygen and that is what makes life possible here," Miss Kelly said.

Miss Kelly led the children to the playground. The children gathered around her, curious to know why she had brought them there.

"Bob, come here and hold this yellow ball," said Miss Kelly, asking Bob to stand in the middle of the playground.

"Bob is the Sun. The Sun is at the centre of the solar system."

She gave the blue ball to Nina.

"Nina, you are the Earth. You have to move around Bob. Children, this is how all the planets revolve around the Sun. We have 365 days in one year because that is the time it takes the Earth to complete one revolution around the Sun."

"Oh! Is that why we have day and night on Earth, Miss?" Mary asked.

Miss Kelly laughed. "No Mary. It is due to the rotation of the Earth that we have day and night. The side of the Earth which faces the Sun has day, while the side which faces away from the Sun has night," Miss Kelly explained.

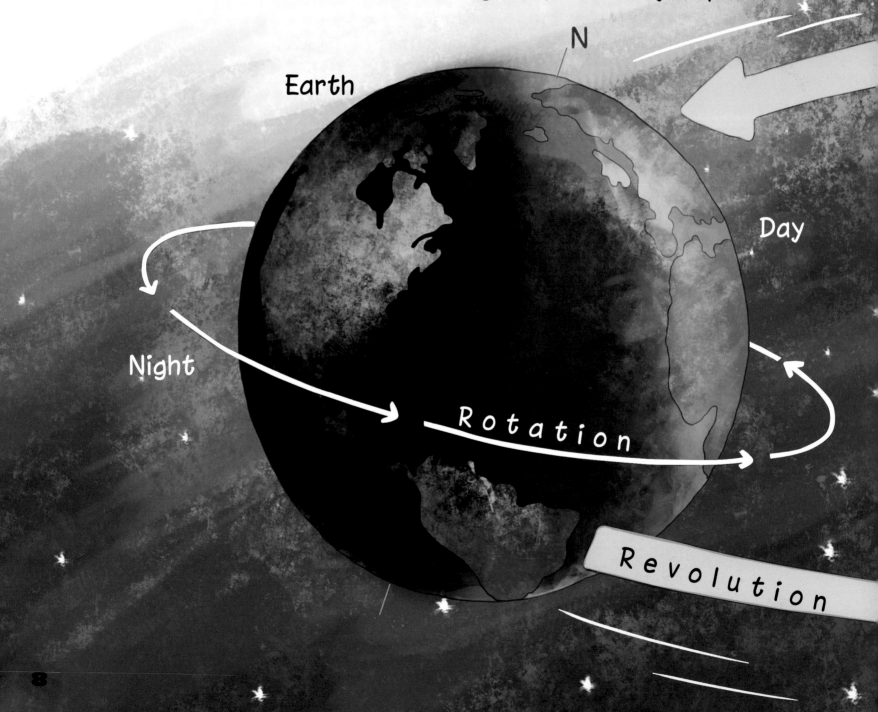

N

Earth

Day

Night

Rotation

Revolution

Revolution

Sun

"Why are there seasons on Earth?" Sam asked.

"We have different seasons because Earth is tilted on its axis. This means that the Earth doesn't rotate in a straight line. When the northern part of the Earth is tilted towards the Sun, it experiences summer. Meanwhile, the southern half which is tilted away from the Sun has winter, and vice versa.

Can anyone name all the seasons?"
Miss Kelly asked.

Nina raised her hand and said,"I can! There are four seasons. They are spring, summer, autumn and winter."

"Very good, Nina!"

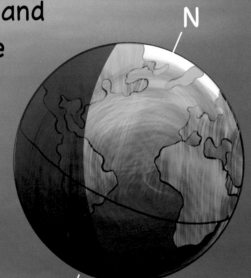

N

Spring

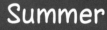

Summer

"But do you know that Earth is in danger?" She asked.

"Why?" Bob asked, shocked.

"The Earth helps us to live. It's friendly. But we don't respect it. We are making the Earth's air, water and land very dirty. We call it pollution. We also harm its best friend, the Ozone layer that protects us from harmful ultravoilet rays of the sun. The Earth is getting sicker day by day. For that reason, the beautiful seasons are losing their flavour too."

Spring

Summer

Autumn

Winter

"Exactly! Human activities are destroying the Earth. The smoke from factories and vehicles pollutes the air. The waste we dump on land and in water harms the environment. The forests are disappearing because we are cutting down trees for our selfish needs. All this is damaging the Earth," Miss Kelly said.

"That is so wrong!" Nina said, sadly.

"If we continue harming our planet, soon humans won't be able to live here. Would you want that to happen?"

"No, Ma'am. We will protect the environment. We love our planet!" the children said in unison.

Ecosystems

The Network of Life!

Om
KIDZ
An imprint of Om Books International

It was a beautiful Sunday morning. Jane and Judy were standing by their bedroom window.

Suddenly, Jane exclaimed, "Look Judy! There is a small mound of sand in the middle of our garden. Where did that come from?"

"It's an anthill. Ants live underground. They make colonies by digging up sand," Judy explained.

"I don't like ants. They crawl around everywhere and even bite. Why do they even exist?" Jane shuddered, thinking about the time when ants had crawled up inside her dress and bitten her.

"Ants, like all other living creatures, are an important part of the ecosystem," explained Judy.

"What is an `ecosystem`, Judy?" Jane asked.

"The plants and animals as well as non-living things that are found in a particular area are referred to as an `ecosystem`. The plants and animals present in an ecosystem are dependent on each other, as well as on their environment for survival. Every ecosystem is based on a very delicate balance. Even the slightest disturbance would affect all the creatures that are a part of it," Judy explained.

"Can you tell me more about ecosystems?" asked Jane.

"There are several types of ecosystems. Some ecosystems exist in a small area such as the ant colonies underneath a rock, a decaying tree trunk, or a pond in your town. Some ecosystems exist in large forms such as a forest or lake. Amazingly, the Earth can be called a huge ecosystem. Do you remember our camping trip last summer?" Judy asked.

"Yes!" Jane replied with a smile.

They had set up a tent in the forest and had so much fun on that trip. A forest is also an ecosystem. It is a large area of land that has a variety of trees, plants and animals. Forests that have warm temperatures and receive a lot of rainfall have a variety of plants and animals.

Forests that are located in the cold and dry regions have less diversity.

"I remember that there were mountains behind the forest. They looked so huge!" Jane chimed in.

"Yes. The mountain ecosystem is barren and rocky with very high peaks. The climate changes according to the height of the mountain, so plants and animals have to adapt to the decreasing temperatures. The plants found in this ecosystem include shrubs, short grasses, mosses and lichens."

"Are there any other types of ecosystems?" Jane questioned.

"Yes!" Judy replied. "We also have the desert ecosystem which experiences extreme temperature changes. Plants such as the cactus are found here. They have thick leaves in order to ensure that there is minimum loss of water for the plant. Animals such as camels that live in deserts have 'humps' to store food for long periods of time."

"Are oceans and seas also a type of ecosystem?"

"Yes. The marine ecosystem is the biggest ecosystem in the world! It covers about 71% of Earth. It includes the five main oceans: the Pacific Ocean, Atlantic Ocean, Indian Ocean, Arctic Ocean and Southern Ocean, as well as many smaller gulfs and bays. Millions of varieties of fish, marine animals, insects and plants are found underwater," replied Judy.

"Judy, how do ecosystems maintain a balance?" asked curious little Jane.

"The plants and animals in an ecosystem are connected to each other through `food chains`. Food chains help in maintaining the balance of an ecosystem. Plants form the first level of every food chain. They are known as producers as they produce their own food with the help of sunlight.

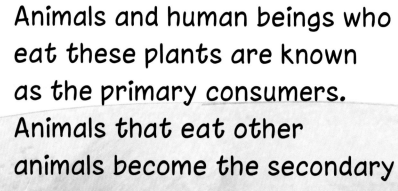

Animals and human beings who eat these plants are known as the primary consumers. Animals that eat other animals become the secondary consumers."

"Earthworms are called decomposers, as they eat the bodies of dead creatures."

"How will an imbalance in the ecosystem affect all the living creatures?" asked Jane.

"Human activities have resulted in a lot of environmental pollution. People are polluting land, water and air by dumping waste. Smoke from factories and vehicles are causing global warming. Forests are being cut to make farmlands and cities, forcing animals out of their homes. These activities disturb the balance of the ecosystems."

"Since all living creatures are connected to each other, they will all be affected by this imbalance!"

Jane was shocked to hear this. "Oh! I don't want the ecosystem to be destroyed. Is there any way that we can help conserve the ecosystem?"

Judy smiled and said, "We can contribute to the conservation of the environment in several ways, such as by planting more trees, using environment-friendly modes of transport and ensuring proper disposal of waste!"

"Let's do our bit too, by planting a tree," said Jane, enthusiastically. And together, the two sisters planted a sapling in the garden while enjoying a beautiful spring morning.

Air, Water and Noise
Pollution

An imprint of Om Books International

Sam was walking down the street towards his friend Ron's place. Since it was a hot and sunny day, the friends had decided to stay in and play video games.

Sam reached Ron's house and found Ron's father in the living room, watching television. Ron's father was watching the news, where the collapse of a landfill was being reported. The news anchor was talking about pollution and its side effects.

Sam had never heard this word before. "What is pollution?" he asked.

"The contamination of the environment with harmful materials is called `pollution'. Pollution affects the Earth as well as plants, animals and humans living on it."

"Let's take a walk down the street to the ice cream parlour. It's too hot and we could all use a popsicle," said Ron's dad, leading the children out of the house.

By now Sam and Ron were both very curious and wanted to know more about pollution. "Dad, are there different types of pollution?" asked Ron.

"Yes. There are four types of pollution—land, air, water and noise pollution. Dumping garbage, throwing waste and littering the streets are all types of land pollution."

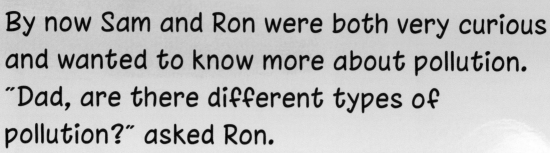

Just then they passed a garbage truck parked by the road. "That's a lot of garbage," Ron exclaimed.

"Yes. Human beings produce a lot of waste every day and this waste is dumped in places called 'landfills'. Waste is produced in both factories and homes. Often, this waste contains a lot of harmful chemicals, which seep into the

water and soil, and affect the health of plants, animals and human beings. These toxins can prove fatal for living beings."

"That sounds scary!" said Sam, shocked.

Just then they walked past the pond. The boys ran towards it as they wanted to play with the ducks. "Where are all the ducks?" Ron asked, surprised.

"I think they all went away because the water is so dirty," said Sam.

"You are right! This is called water pollution. When water in ponds, lakes, rivers or oceans gets dirty, the plants and animals living there are affected," Ron's dad explained.

"What are the causes of water pollution?" asked Sam.

"Dumping of waste into water bodies is one of the main causes of water pollution. People who visit lakes and ponds litter it with food packets and bottles. Even incidents such as oil spills can cause water pollution. Tankers carrying oil may leak or get into accidents leading to oil spills in oceans. This layer of oil spilled on the surface of the ocean water doesn't allow sunlight and oxygen to reach the marine plants and animals, and may prove fatal for them."

While walking ahead, the boys spotted thick smoke coming out of a tall chimney. "What is that?" asked Ron.

"Those are the fumes from the factory. These fumes cause air pollution. They pollute the air we breathe and this is harmful for our health. Exhaust fumes from vehicles and dust particles also cause air pollution," Ron's dad explained.

Just then a car passed by, blaring its horn. The children instantly covered their ears to block out the loud noise.

"I wish people didn't honk so loudly," Sam said angrily.

"This is another form of pollution-noise pollution. The loud noises from vehicles, construction sites, machines and loud speakers can cause a lot of damage to our health. It can result in problems such as hearing loss, stress and even sleep disturbance," Ron's dad explained.

"Dad, how can we to stop pollution?" asked Ron.

"The best way to control pollution is by following the three R's–Reduce, Reuse and Recycle. We should dispose garbage in an eco-friendly manner.
We should also plant more trees, which will help the environment," Ron's dad replied.

Finally, they reached the ice cream parlour and Ron's dad got them orange popsicles. While relishing his popsicle, Ron spotted a nursery on the opposite side of the street. "Dad, can we buy some saplings from the nursery?" Ron asked.

"That's a great idea! We can plant them in the garden, today itself. This way, we too can help save planet Earth," Sam declared, happily.

"Sure and you can call yourselves Captain Planet!" said Ron's dad, smiling at the kids.

Global Warming

An imprint of Om Books International

Bob decided to host a party. He kept the party theme – superheroes. His friends were excited about the news. They dressed up in their favourite superhero costumes and arrived at his house. It was a lively get together of friends. John was Zatana, Nina was Wonder Woman and Kim was Supergirl.

"Who are you, Bob?" Nina asked Bob.

"I am Captain Earth, I am here to protect our planet," answered Bob.

Bob's Dad overheard him and asked, "How do you know our planet needs help?"

"Dad, Miss Victoria told us that our Earth is boiling in anger. We are harming the Earth," exclaimed Bob.

"Yes, it is true! Earth is getting hotter day by day. It is called 'global warming'," added Bob's Dad.

Bob grew curious, "Dad, can you explain a bit more?"

"Our atmosphere is a layer of gases surrounding the Earth that prevents temperatures from rising too high or getting too cold. The atmosphere also absorbs harmful ultra violet radiations from reaching us," explained Bob's Dad.

Nina asked, "Isn't that the greenhouse effect?"

"You are right Nina, that's the natural greenhouse effect. But, we are talking about the bad greenhouse effect here," Bob's Dad said.

"What is the bad greenhouse effect?" John asked.

"Greenhouse gases include water vapour, carbon dioxide, methane and nitrous oxide. Except water vapour, other three gases are mostly released by people." Bob's Dad explained. "Our planet is getting warmer as some greenhouse gases in the atmosphere are increasing due to people's activities."

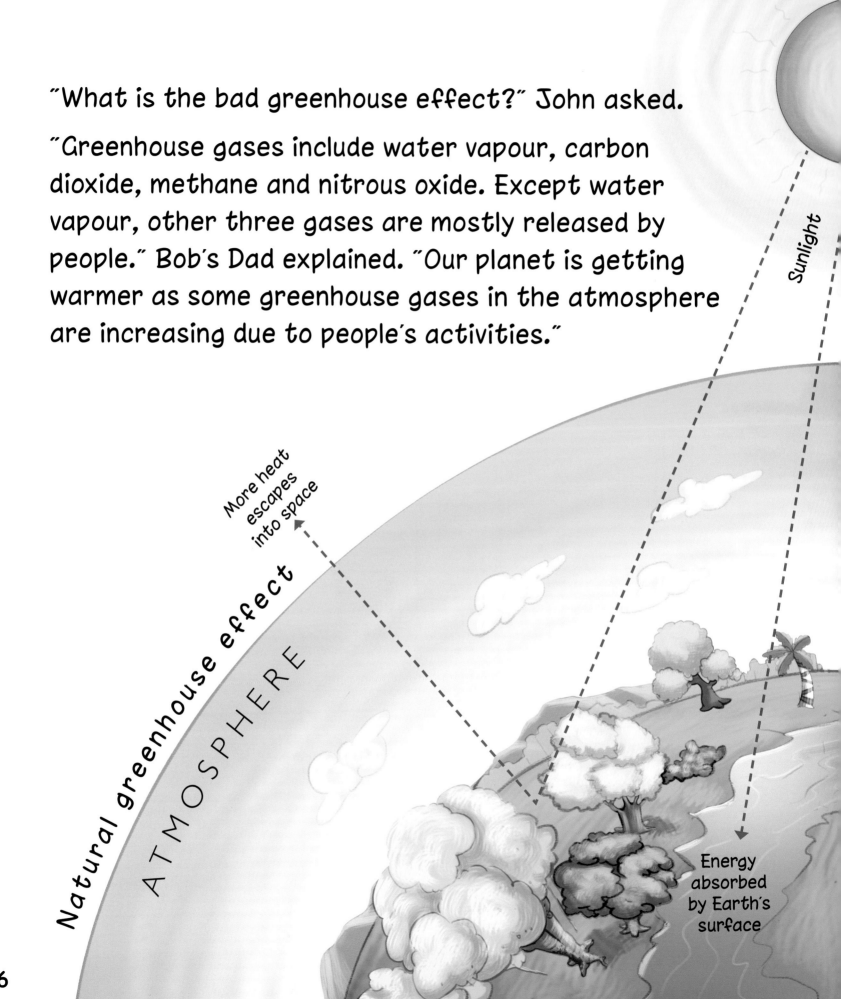

Sunlight

More heat escapes into space

Natural greenhouse effect

ATMOSPHERE

Energy absorbed by Earth's surface

"Really?" Nina cried out. "Are we responsible for our planet heating up?"

"That's true Nina," answered Bob's Dad. "Did you know that the 11 hottest years recorded in the last 100 years have all happened after 1995. The world is definitely getting warmer."

Sunlight

Greenhouse gases trap heat in the atmosphere so less heat escapes into space

Harmful greenhouse effect

ATMOSPHERE

Energy absorbed by Earth's surface

"Burning of fossil fuels like oil, coal, natural gas by people are the real culprits," said Bob's Dad.

"Miss Victoria told us that fossil fuels contain carbon. When they are burnt, this carbon is released and gets combined with oxygen in the atmosphere to form carbon dioxide," Kim added.

"There is another reason for global warming," Bob's Dad said.

The children looked surprised. "There is more to our planet's woes?" said Nina

"People are cutting forests to make land available for other uses. This has led to larger carbon dioxide emissions in the atmosphere," Bob's Dad continued.

"Is that because trees and other plants absorb a lot of carbon dioxide on the Earth's surface?" asked Bob.

"That's right Bob, but the forest cover across the world is reducing every passing day. About 36 football fields worth of trees are lost every minute," answered Bob's Dad.

9

"The Earth is falling sick again and again because of global warming. Glaciers are disappearing. Sea levels are rising. Heat waves, storms and floods are becoming more extreme. What's more, insects are emerging sooner and flowers are blooming earlier. Some birds are laying eggs before they're expected and bears have stopped hibernating," Bob's Dad described.

"Because of global warming there is a rise in water temperature in the oceans too. This has led to the

waning of coral reefs that are home
to many living creatures. Sea turtles
depend on beaches to lay their eggs, but
many beaches are disappearing, thanks
to rising sea levels. Fish stocks are
falling as their staple food-plankton and
krill-is dying due to high temperature in
the oceans," Bob's Dad added.

"Dad, can we do anything to save our planet?" Bob asked.

"Well, if you reduce your `carbon footprint` you can contribute to Earth's welfare," his Dad answered.

"What's a carbon footprint?" Kim asked, wondering.

"Your carbon footprint is the amount of carbon dioxide released into the air because of your own energy needs. You need transportation, electricity, food, clothing, and other things. Your choices can make a difference," Bob's Dad described.

"How can we reduce our carbon footprint?" John questioned.

"When you use electricity in your home – while watching TV or playing video games – you are creating carbon dioxide. How? That's because the power plants burn coal to produce electricity, releasing the most amount of carbon dioxide.

Gas heating used to warm up your home during winter is the second largest source of carbon dioxide emissions.

For every pound of trash you throw away, you create one pound of greenhouse gases. As it decays, trash creates carbon dioxide and methane.

And...we all know that cars contribute to our carbon footprint!" Exclaimed Bob's dad.

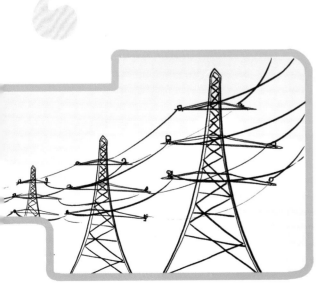

"Captain Earth you were right, our planet needs help," Kim cried out helplessly. "What should we do?"

Captain Earth already had a plan to save planet Earth. Kim, Nina and John began creating a to-do list to reduce their carbon footprint.

- Plant trees

- Try composting

- Reduce, reuse, recycle

- Turn off the lights, unplug electronics

- Only do full loads of laundry or put clothes to dry

- Take shorter showers, turn off the tap

- Skip the car ride and use your bike, or take the bus or walk.

- Don't buy bottled water; drink tap water, and filter it if you like

- Use less paper

- Eat green vegetables and fruits; less meat and processed food.

All the superheroes along with Captain Earth geared up to get, set, go. They began by planting trees in the garden.

They also promised to reduce their carbon footprint in every possible way.

It is never too late to help!

Endangered Species

An imprint of Om Books International

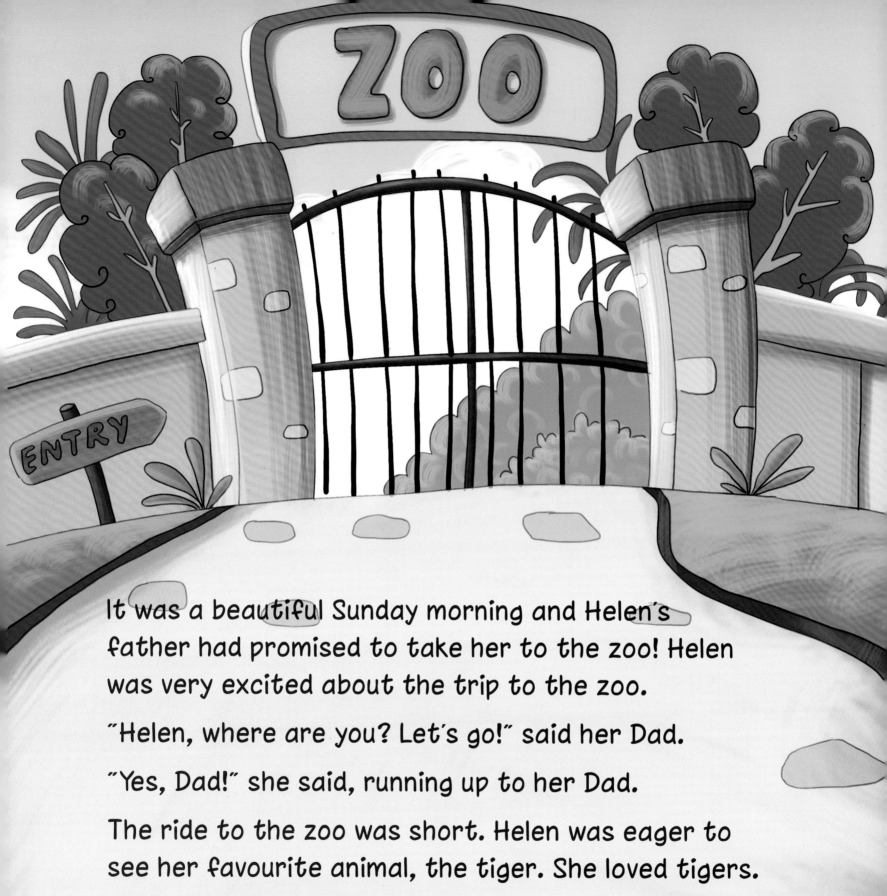

It was a beautiful Sunday morning and Helen's father had promised to take her to the zoo! Helen was very excited about the trip to the zoo.

"Helen, where are you? Let's go!" said her Dad.

"Yes, Dad!" she said, running up to her Dad.

The ride to the zoo was short. Helen was eager to see her favourite animal, the tiger. She loved tigers.

"Dad, where are the tigers? I want to see them up-close," she said.

"Of course! Come, let me take you there. Helen, do you know why tigers are a special attraction in this zoo?" asked Dad.

"Why?" asked Helen, curious.

"It is because they are an endangered species and hence, rare." said Dad.

"What are endangered species?" she asked.

"An endangered species is any type of flora or fauna that is in danger of disappearing forever. When a type of plant or animal dies out completely, it is referred to as an `extinct species'."

This scared Helen. She didn't want the tigers to disappear from Earth!

"Why do animals and plants become endangered?" asked Helen.

"Animals and plants depend on each other and their surroundings for survival. This is called their `habitat'. Plants and animals become endangered when their natural habitat changes or disappears. Climate change, natural disasters and human activities also endanger plants and animals," said Dad.

4

"How do human activities endanger animals?"

"Human activities such as cutting down trees to make roads, houses and factories harm the environment. People also dirty the rivers and lakes and land with garbage, which leads to pollution. Also, the smoke from factories and vehicles pollutes the air. All these things have endangered the lives of animals and plants."

"There's the tiger! He looks so scary and ferocious!" Helen exclaimed as they finally reached the enclosure where the tigers were kept.

"The tiger is the largest wild cat in the world. Tigers have become endangered because humans kill them for their fur and other body parts. Human activities such as cutting trees have resulted in the destruction of the natural habitat of the tigers."

They walked on to the next enclosure, which had three elephants.

"Oh! They are so huge and look at their long noses!" she said, clapping her hands gleefully.

"An adult elephant's trunk can be about seven feet long! It's actually an elongated nose but elephants can even drink from it," informed her Dad.

"That sounds so cool! Are elephants also in danger of becoming extinct?" Helen asked.

"Yes. Elephants are endangered because their habitat is slowly disappearing due to humans cutting down forests. Also, elephants are hunted for their ivory tusks."

Next they came across a huge aquarium. Helen ran forward and pressed her nose to the glass structure.

"There are so many colourful fish inside. It's so beautiful! Dad, are fish also endangered?" she asked.

"Yes dear! The blue whale is the largest mammal in the world and lives in oceans and seas," said Dad, showing her a sculpture of the blue whale displayed on a platform.

"The population of blue whales is decreasing at an alarming rate. About 90% of the blue whales have disappeared in the past hundred years. They are hunted for their blubber and oil."

As Helen and her Dad walked ahead, they heard a huge commotion. It was coming from the next enclosure! Helen ran ahead to see what was happening.

"There are so many birds here! Look Dad, they are making so much noise while flying around. Does the list of endangered species include birds as well?" asked Helen.

"Yes. There are many birds which are at the risk of extinction, such as the bald eagle, the whooping crane, the emperor penguin and several others. Birds are hunted for their feathers and other body parts. The population of birds is decreasing due to environmental pollution. Climate change also poses a huge threat to the life of birds and animals," replied her Dad.

"What can we do to save these creatures, Dad?" Helen asked.

"The best thing that humans can do is respect Mother Nature. We must stop hunting these creatures for selfish reasons. We can also come up with ways to reduce pollution, recycle trash and stop cutting trees."

"I promise to never hurt planet Earth. I will also try to raise awareness about the endangered species so that more people can join the cause of saving planet Earth and the creatures living on it," said Helen.

3R's

Reduce

Recycle

Reuse

Om
KIDZ

An imprint of Om Books International

One day, Sam was playing with his favourite toy robot, who he had named Rob. Rob was metallic and shiny, and could even walk!

Suddenly Sam cried out, "Oh no!"

"What's wrong, Sam?" his Mom asked, rushing to his side.

"I was playing with Rob and his arm came off," Sam cried helplessly.

Sam's mother tried her best to fix the robot's arm, but she couldn't. Sam became very sad.

"Will you throw him away, Mom? What will happen to Rob now?"

"Don't worry Sam. Rob will be okay. Do you know what happens to things when you throw them away in the trash?" Mom asked.

"The trash man comes and takes them away in his huge smelly truck. Will Rob also live in the smelly truck now?" Sam asked, sadly. He didn't want to think about Rob living there.

"No dear, Rob won't live in the smelly truck. He will be recycled and turned into something new, maybe another robot."

"Woah! How will that happen?" Rob asked.

"Rob is made of metal and all metal objects can be recycled. Now tell me, how many dustbins do we have?"

Sam thought and quickly replied, "Three!"

"That is correct! We throw all our metal and glass trash into one dustbin. The kitchen waste, such as vegetable peels and stale food, goes into another dustbin. The third dustbin is for plastic waste," she explained.

"Why do we separate trash into three bins?"

Food & Garden Waste

Recyclable Waste like paper and metal

6

"There are three types of waste and they are disposed of in different ways. The kitchen waste is dumped into a huge hole called a 'landfill'. Over time, this waste decomposes," his Mom explained.

Non-recyclable Waste

"What do you mean by `decompose`?" Sam questioned.

"It means that this waste decays and mixes with the soil. Metal trash does not decay. It needs to be recycled at the Recycling Centre."

"Is that where Rob will go?"

"Yes, Sam. Rob will be taken to the Recycling Centre. There, he will be melted. The melted metal will then be used to make another object."

"That is so cool, Mom! Rob will change into something else. Just like the transformers!" Sam was very pleased to hear this.

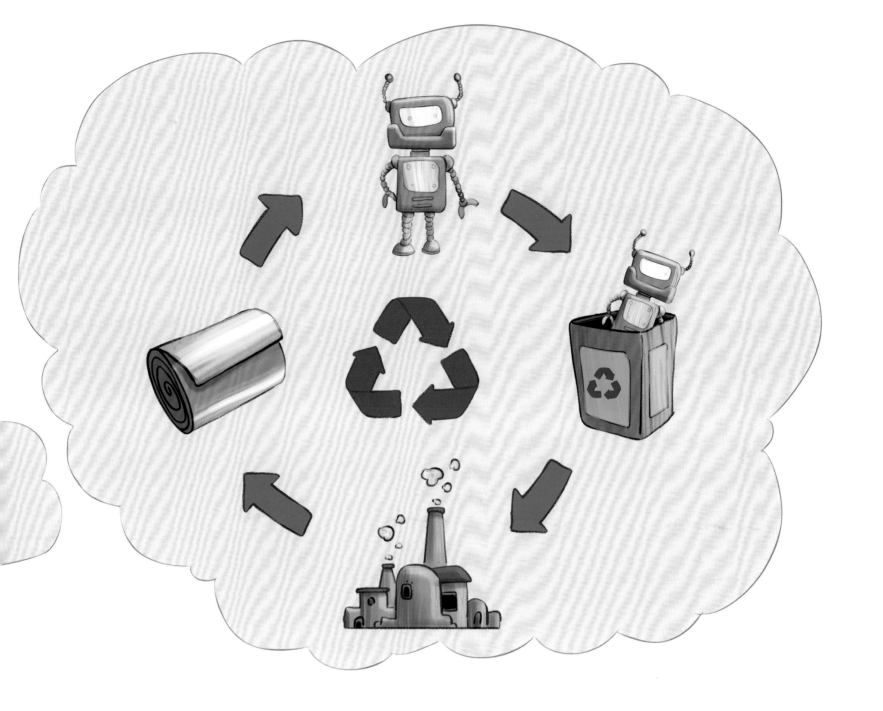

"But Mom, why do we need to recycle?" asked Sam.

"Humans have harmed nature a lot due to their actions. Our planet is in danger! To save our planet we all must learn to practice the `three R's`," Mom explained.

"What are the `three R's`?" Sam asked.

"The first R stands for reduce. This means that we should cut down our usage and buy less. There are many ways in which we can reduce. We should avoid wasting food and only take as much as we can eat. We should switch off the lights and fans when they are not in use. Instead of travelling by car, we should walk or cycle."

"We should reduce our water consumption in little ways, such as by making sure that the taps are not left running while we brush our teeth."

"Mom, what do you mean by re-use?"

"Instead of throwing away things, we should try to find ways to re-use them."

Sam was very confused. "How can I re-use things?"

"It is very simple," Mom replied. "You can make paper bags and envelopes out of newspapers instead of

throwing them away. Tell me, what do you do with empty cold drink bottles?" Mom asked.

"I throw them away."

"Instead of throwing them away, you can re-use the bottles. You can use them as flower pots or you can use them to make birdhouses. In this way, you will produce less waste and help the environment."

"And I already know what recycling means. Now I understand what the three R's stand for. I too can help in saving the Earth now, just like a superhero!" Sam said joyously.

Just then, he heard the horn of the trash truck. He ran outside and handed over Rob to the trash man.

"Yay! Now, Rob will also become a transformer!"